SNEAKY PRESS

©Copyright 2022
Pauline Malkoun

A catalogue record for this work is available from the National Library of Australia.

ISBN 9781922641274

Sneaky Press is the imprint of Sneaky Universe.
www.sneakyuniverse.com
First published in 2022

Sneaky Press
Melbourne, Australia.

The Book
of
Random Space Facts

Sneaky Press

Contents

Space Firsts

The first rocket reached space in 1942.

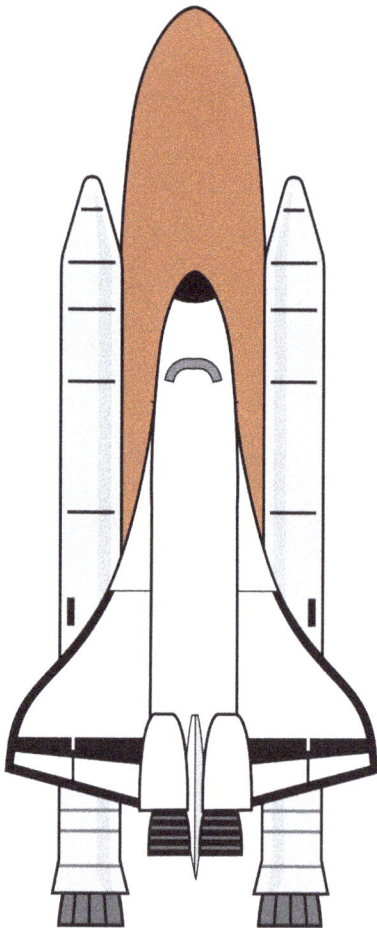

Fruit flies were sent to space in 1947.

The first mammal sent to space was a monkey called Albert II in 1949.

In 1957, a dog named Laika orbited the earth.

Luna 1, a Russian unmanned space craft, crash landed on the moon in 1959.

Russian cosmonaut Yuri Gagarin is the first human in space on April 12, 1961 aboard Vostok 1. He spent 108 minutes there and orbited the earth once.

On February 20, 1962, American astronaut John Glenn orbited the Earth 3 times aboard Friendship 7. He spent four hours and 55 minutes in space.

The first woman in space was Russian cosmonaut Valentina Vladimirovna Tereshkova on June 16, 1963. She spent 70 hours in space and orbited the earth 48 times.

Apollo 8 astronauts Frank Borman, Jim Lovell and Bill Anders become the first humans to orbit the moon on December 24, 1968

On March 18, 1965, Russian cosmonaut Alexei Leonov is the first man to walk in space.

Astronauts Neil Armstrong, Buzz Aldrin Jr. and Michael Collins were the first humans to land on the moon on July 20, 1969. Armstrong and Aldrin are the first to walk on the moon.

On April 28, 2001, Dennis Tito becomes the first "space tourist," paying $20 million to ride on a Russian rocket to the International Space Station (ISS).

On August 2, 2020, SpaceX's Crew Dragon spacecraft splashed down into the Gulf of Mexico. This is the first time in history that a commercially developed spacecraft carried humans into Earth's orbit.

Size in Space

Pluto
diameter
2374 km

The Moon
diameter
3474 km

Mercury
diameter
4879 km

Mars
diameter
6771 km

Venus
diameter
12 104 km

Earth
diameter
12 742 km

Neptune
diameter
49 244 km

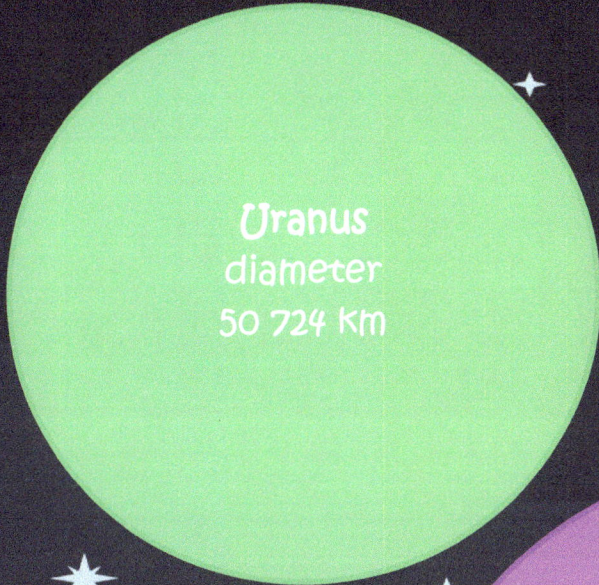

Uranus
diameter
50 724 km

Saturn
diameter
116 464 km

Size in Space

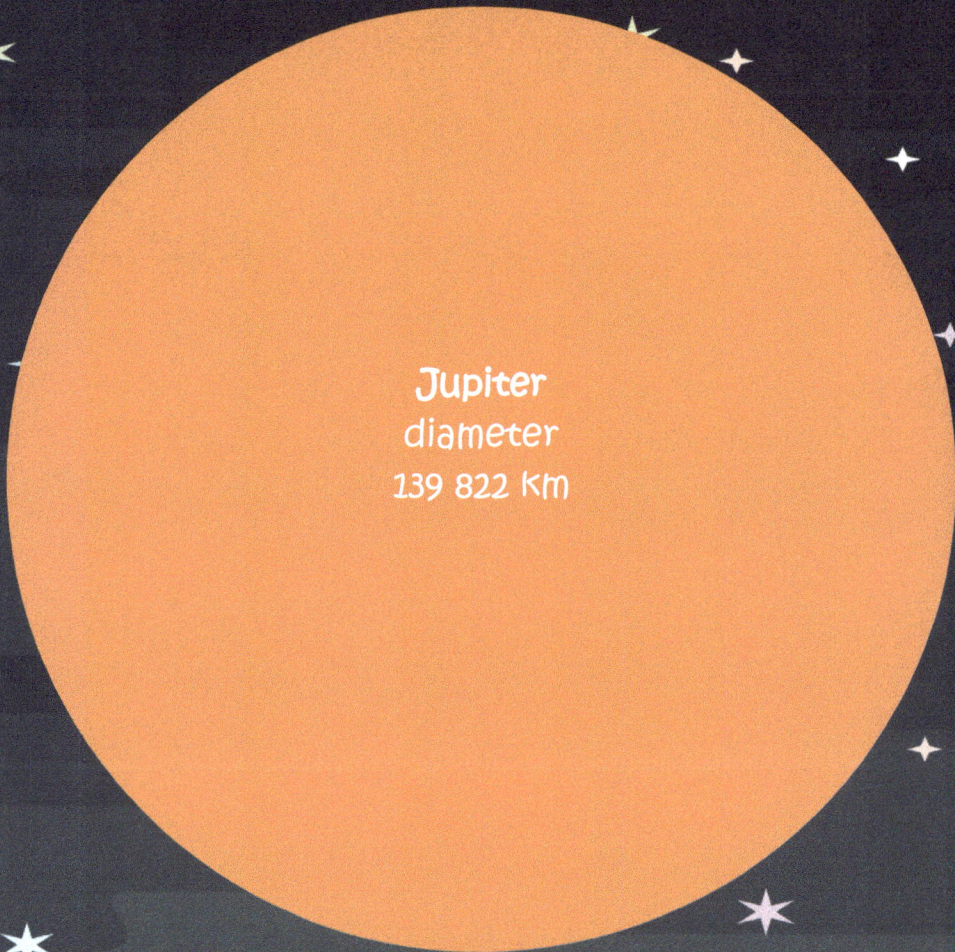

Jupiter
diameter
139 822 Km

The Sun
diameter
1.391016 million km

The Sun accounts for
almost all the mass in
our solar system
(99.86%).

Random Facts about Mercury

Mercury is named after the Roman god of merchants and travellers.

You would weigh 62% less on Mercury than on Earth.

Mercury does not have any moons or rings.

A Mercury day is equivalent to 176 Earth days.

The Mariner 10 was the first spacecraft to visit Mercury in 1974.

Mercury is the second hottest planet.

It's not known who discovered Mercury.

A Mercury year takes 88 Earth days.

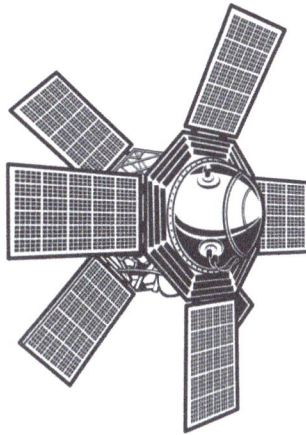

Random Facts about Venus

Venus is named after the Roman goddess of love.

A Venus year takes 225 Earth days.

A day on Venus is the equivalent of 117 Earth days.

The surface
temperature on
Venus can reach
471 °C making it the
hottest planet in
our solar system.

Venus does not have
any Moons.

Venus is the second
brightest object in
the night sky.

Venus rotates in
the opposite
direction to most
of the other
planets.

Random Facts about Mars

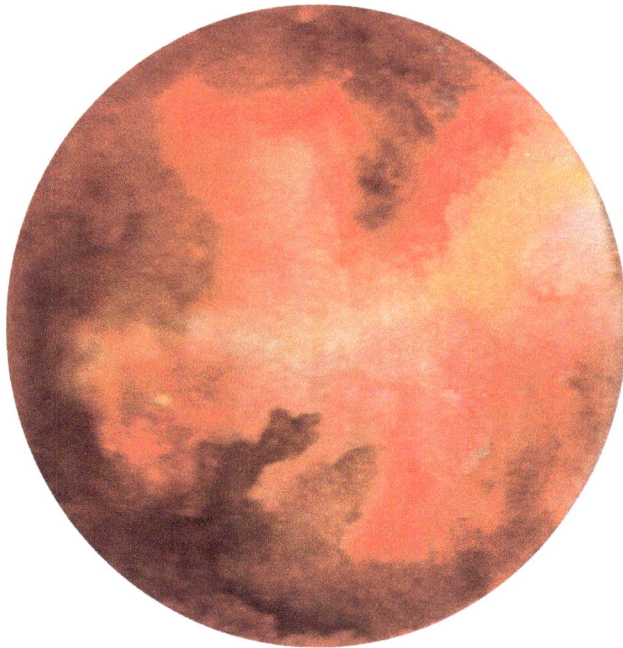

Mars is named after the Roman god of war.

Mars has two moons, Phobos and Deimos.

Only 18 missions out of 40 to Mars have been successful.

There are signs of liquid water on Mars.

Sunset on Mars is blue.

Mars has the largest dust storms in our solar system. They can last months and cover the entire planet.

Mars is home to Olympus Mons, the tallest mountain in the solar system.

Random Facts about Jupiter

Jupiter is named after the Roman king of all gods, he is also the god of light.

Eight space craft have visited Jupiter.

A Jupiter day is the equivalent of 9 Earth hours and 55 minutes – the shortest in our solar system.

Jupiter's Great Red Spot is a storm that has raged for at least 350 years. It is so large that three Earths could fit inside it.

Jupiter emits more energy than it receives from the Sun.

Jupiter orbits the Sun once every 11.8 Earth years.

Jupiter has clouds made mostly out of ammonia crystals and sulphur.

Jupiter has 79 known moons including the largest moon in our solar system, Ganymede.

Jupiter is the fourth brightest object in our solar system.

Random Facts about Saturn

Saturn is named after the Roman god of agriculture.

Saturn can be seen without a telescope in the night sky.

Saturn orbits the Sun once every 29.4 Earth years.

Four spacecraft have visited Saturn.

Saturn has the most extensive rings in the solar system made mostly of chunks of ice and dust. The rings stretch out more than 120,700 km from the planet.

Saturn is the flattest planet.

Saturn is composed mostly of hydrogen.

If you were to drive a car on one of Saturn's rings, at the speed of 100 km/h, it would take over 14 weeks to finish one lap.

H H

Saturn has 150 moons and smaller moonlets.

Random Facts about Uranus

Uranus is named after the Roman god of the sky.

Uranus has 27 moons.

A day on Uranus is the equivalent of 17 hours and 14 minutes on Earth.

Uranus makes one trip around the Sun every 84 Earth years.

Only one spacecraft, the Voyager 2, has ever flown by Uranus in 1986.

Uranus has two sets of very thin dark coloured rings.

Uranus is the coldest planet with minimum recorded atmospheric temperatures of -224 degrees Celsius.

Random Facts about Neptune

Neptune is named after the Roman god of the sea.

Neptune has 14 moons.

Neptune's' atmosphere is mainly composed of hydrogen and helium, with some methane.

Neptune spins on its axis very quickly.

Only one spacecraft, the Voyager 2, has ever flown by Neptune in 1989.

Neptune has a very thin collection of rings.

Neptune has highspeed winds whipping around the planet at up 600 meters per second.

Random Facts about Pluto

Pluto is named after the Roman god of the underworld.

Pluto was reclassified from a planet to a dwarf planet in 2006.

Pluto is smaller than Earth's moon.

Pluto has five known moons.

Pluto is not the only dwarf planet in our system. There are four other dwarf planets: Ceres, Haumea, Makemake and Eris.

Pluto has an elliptical orbit and is at times closer to the sun than Neptune.

The only spacecraft to fly by Pluto was New Horizons in 2015.

Random Facts about the Moon

The moon is the only natural satellite that orbits the Earth.

The moon does not have any atmosphere.

The Moon is in a synchronous orbit around the Earth. This means that we always see the same side of the Moon.

The gravity on the moon is 83% less than on Earth. This means that if there was a pool on the Moon, swimmers could jump out of the water like dolphins, launching themselves more than one metre high.

There is at least one solar eclipse every 18 months. A solar eclipse occurs when the Moon passes right in front of the Sun, and it casts its shadow on Earth.

The Moon does not have a dark side. The side that we don't ever see is illuminated by the Sun as often as the side that we do see.

A Blue Moon is not really blue. It is the name for the second full moon that happens in one month usually once every 2-3 years.

Due to the lack of atmosphere, footprints on the moon will stay there for 100 million years.

The moon is 384,402 km away from Earth.

There are at least two lunar eclipses every year. There can be up to four. A Lunar eclipse happens when the Moon passes into the Earth's shadow, blocking the sunlight that usually falls on the Moon. During a lunar eclipse, we still see the Moon, but it has a weak reddish tint.

Random Facts about Galaxies

The Milky Way contains between 100 – 400 billion stars.

Our galaxy, The Milky Way is approximately 13.6 billion years old.

There are 4 main types of galaxies: Elliptical, Normal Spiral, Barred Spiral and Irregular. Our galaxy, the Milky Way is a Barred Spiral galaxy.

The Andromeda Galaxy is our neighbour, the closest galaxy to ours.

It is thought that there are over 500 billion galaxies in the universe!

Random Facts about Asteroids and Comets

It is thought that there are over a million asteroids in space at the moment.

Asteroids come in a variety of sizes. They can be as small as a few metres to hundreds of kilometres wide.

Comets are like snowballs in space. They are made of frozen water and gas, rock and dust.

Asteroids are separated by at least several kilometres so avoiding them when flying through space is not difficult.

The Kuiper belt is a disc shaped region of comets, asteroids and dwarf planets. It is though that there are thousands of bodies larger than 100km and trillions of comets in it.

Halley's Comet is the earliest recorded comet, with the first recorded observation in ancient China in 240 B. C. E. It orbits the Sun every 75 years.

A comet's tail, which can be millions of kilometres long, appears when it gets close enough to Sun and starts to melt.

The nucleus of a comet is usually smaller than 10km, but as they approach the sun, the frozen gases evaporate and then the nucleus can expand to over 80 000km.

Random Space Station Facts

Usually there are seven people living and working on the International Space Station.

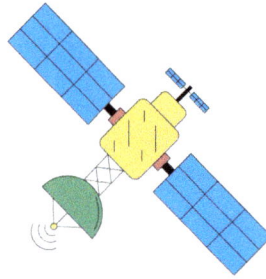

The International Space Station is operated by five space agencies and 15 countries.

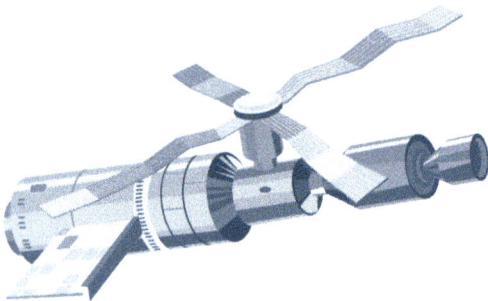

The International Space station has been continuously operating since November 2000.

Eight spaceships can be connected to the space station if needed.

In 24 hours, the space station orbits Earth 16 times.

The space station is 109 metres long.

It takes only four hours to reach the space station from Earth for some space craft.

There are four different cargo spacecraft that deliver supplies to the space station: Northrop Grumman's Cygnus, SpaceX's Dragon, JAXA's HTV, and the Russian Progress.

The space station travels the equivalent distance to the Moon and back each day.

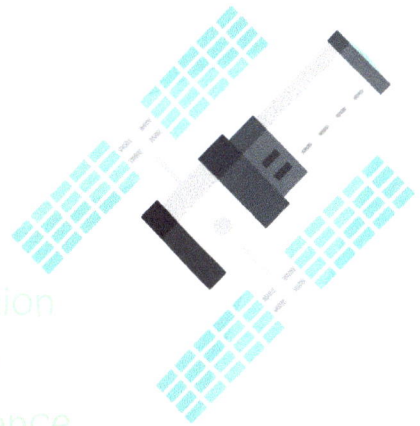

All astronauts on the space station have to exercise for at least two hours each day to stop muscle and bone loss.

Random Facts about Space

Each space shuttle launch costs $450 million.

To break free of Earth's gravity, a space craft has to travel at a speed of approximately 24,000 kilometres per hour.

Due to the lack of gravity, if you cry in space, your tears will not fall down.

Normal pens do not work in space due to the lack of gravity.

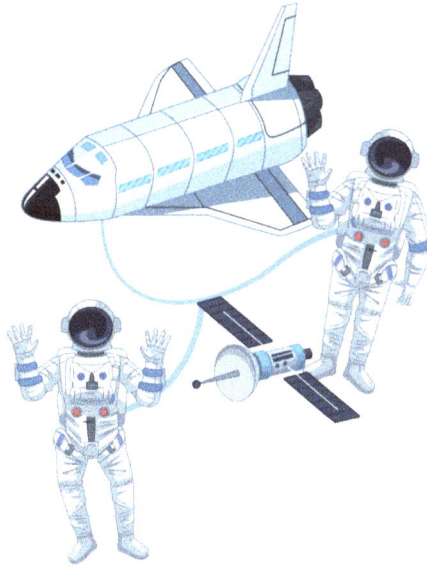

The Sun travels around the galaxy once every 200 million years.

Due to the lack of atmosphere, space is completely silent. Sound waves has no way to travel through the air. Astronauts use radios to communicate because radio waves do not need atmosphere to travel.

A space shuttle needs 1.9 million litres of fuel to launch into space. That's enough fuel to fill up 42,000 cars!

More Random Facts about Space

The first food eaten in space was applesauce.

Due to the lack of gravity, people are 5cm taller in space.

You can't burp in space because the lack of gravity does not allow air in the stomach to rise up from the food that has been eaten.

The first artificial satellite in space was Sputnik. It was launched in October 1957.

The first soft-drink consumed in space was Coca-cola.

Stars appear to twinkle because the light is disrupted as it passes through the Earth's atmosphere.

The origins of the word astronaut translates to "star sailor".

Space Idioms

To shoot for the stars means to aim high, often used when trying to do something challenging or new.

To have stars in your eyes means to be hopeful about the good things that could happen in the future.

To thank your lucky stars means to be grateful for some good luck.

If someone is living on another planet, they are not alert or aware of what is happening around them, or they are not being realistic.

If something is not rocket science, it means that it is does not require intelligence or extraordinary skill.

If you are over the moon, you are very happy about something.

If something is written in the stars it means that something is definitely going to happen.

Everything under the sun means everything that a person could imagine.

When you come back down to Earth, you are doing something boring after having done something exciting.

If something happened many moons ago, it happened a long time ago.

Space Jokes

Why do aliens always spill their tea? Because they have flying saucers.

What is the slowest species in the galaxy? Snailiens.

What did Venus say to Saturn? "Give me a ring sometime."

What kind of money do aliens use? Star bucks.

What is an alien pet called?
An extra furrestrial.

Why did the sun go to school every day? To make sure it got brighter.

What do you call a lazy person in space?
A procrastonaut.

What do you get when you cross a lamb and a space shuttle?
A space sheep!

Why couldn't the astronaut book a room on the moon? Because it was full!

Why doesn't anybody trust the man on the moon?
Because he has a dark side.

Other titles in the

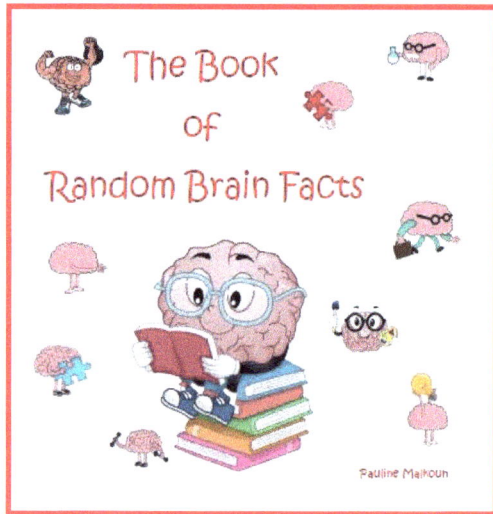

The Book
of
Random Brain Facts

Pauline Malkoun

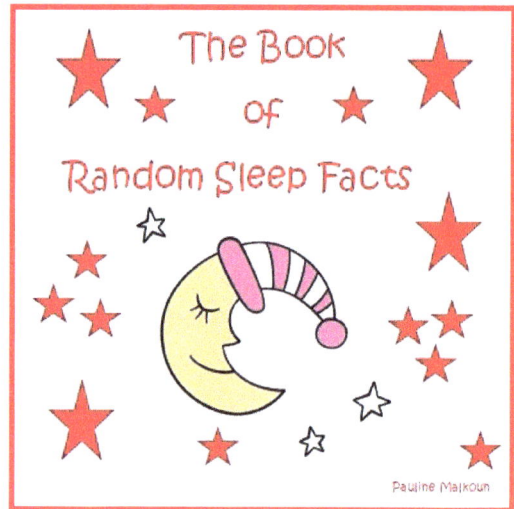

The Book
of
Random Sleep Facts

Pauline Malkoun

Random Facts Series

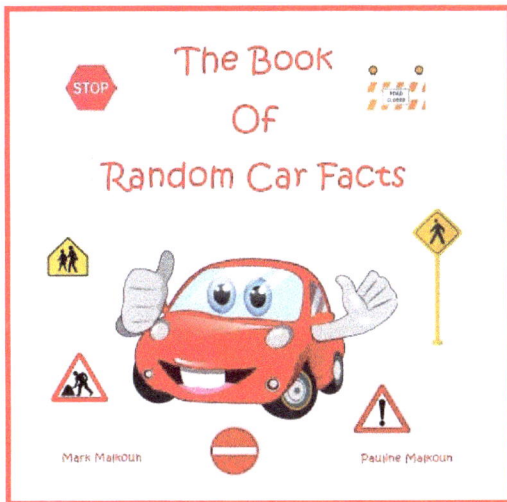

The Book
Of
Random Car Facts

STOP

Mark Malkoun

Pauline Malkoun

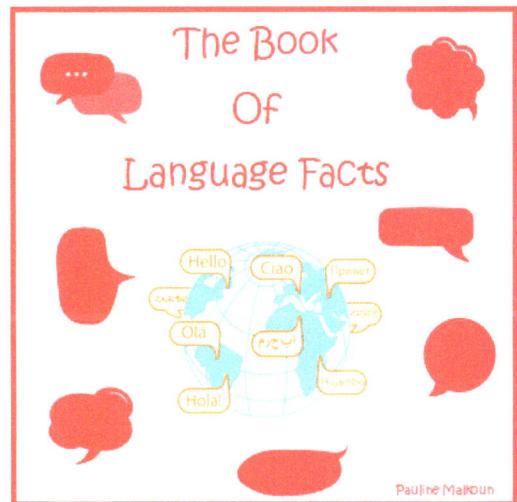

The Book
Of
Language Facts

Hello
Ciao
Ola
Hola!

Pauline Malkoun

www.ingramcontent.com/pod-product-compliance
Lightning Source LLC
Chambersburg PA
CBHW042334030426
42335CB00027B/3332